ASTEROIDS, COMETS, and METEORS

Robin Kerrod

Lerner Publications Company • Minneapolis

This edition published in 2000

Lerner Publications Company
A division of Lerner Publishing Group
241 First Avenue North, Minneapolis MN 55401 U.S.A.

Website address: www.lernerbooks.com

Library of Congress Cataloging-in-Publication Data

Kerrod, Robin.
    Asteroids, comets, and meteors / Robin Kerrod.
        p. cm. – (Planet library)
    Includes index.
    Summary: Introduces asteroids, comets, and meteors,
including their origin, composition, orbits, and effects on
Earth and other bodies in the solar system.
    ISBN 0-8225-3905-5 (lib. bdg.)
    1. Asteroids—Juvenile literature. 2. Comets—Juvenile
literature. 3. Meteors—Juvenile literature. I. Title. II
Series: Kerrod, Robin. Planet library.
QB651.K44   2000
523.5—dc21   99-18759

Printed in Singapore by Star Standard Industries [PTE] Ltd
Bound in the United States of America
2 3 4 5 6 7 – OS – 07 06 05 04 03 02

Contents

Introduction

Nearly 5 billion years ago, the solar system began to form from a great cloud of gas and dust. Most of this material formed the Sun, the star at the center of the solar system. The rest of the material continued circling the Sun, eventually forming the nine planets and their moons. After these larger bodies formed, many smaller lumps of matter were left over. These lumps are what we call asteroids, comets, and meteors. They were scattered throughout space between the planets and at the edge of the solar system. Many of the smaller leftover particles rained down on the newly formed planets and moons.

Asteroids, comets, and meteors are made up mostly of rock, ice, and metal. Asteroids and meteors are mostly rock, but some also contain metal. Comets are a mixture of rock, ice, and dust. Scientists sometimes call them "dirty snowballs." Scientists study the ancient materials in these small worlds to learn more about how our solar system began.

Below: The 1965 comet Ikeya-Seki
showed up brilliantly in the western
sky after sunset. It was one of the
brightest comets of the 20th century.
The orange light to the left of the
comet comes from the Sun.

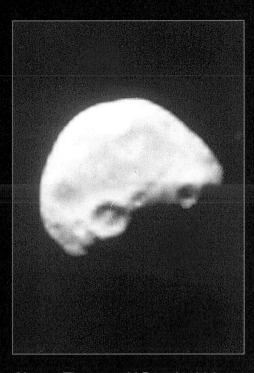

Above: The asteroid Dactyl, which
was discovered by the space probe
Galileo on its way to Jupiter,
measures about 1 mile (1.6 km)
across.

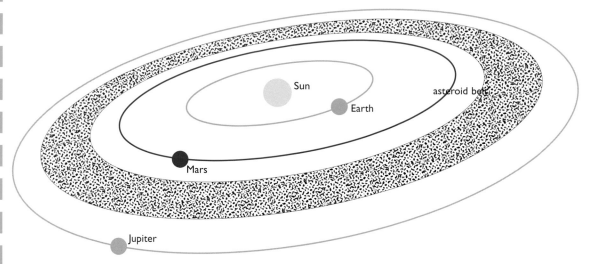

Most asteroids are found in a broad ring, or belt, between the orbits of Mars and Jupiter.

The Asteroids

Most asteroids circle the Sun in a wide band between the orbits of Mars and Jupiter. Scientists have identified more than 7,000 of these rocky bodies, but as many as a million more may exist.

The asteroids are the largest of the small worlds that orbit the Sun. But even Ceres, the biggest asteroid, is less than one-third the size of the Moon and about half the size of the smallest planet, Pluto. Most asteroids, however, are much smaller than Ceres. Many have a diameter, or distance across, of a few tens of miles.

DISCOVERING ASTEROIDS

In 1801, an Italian astronomer named Giuseppe Piazzi discovered the first asteroid. An astronomer is a scientist who studies outer space. Piazzi thought the object was a comet, but it proved to be a small body orbiting about midway

Below: Even the three biggest asteroids—Ceres, Pallas, and Vesta—are very small compared to Earth.

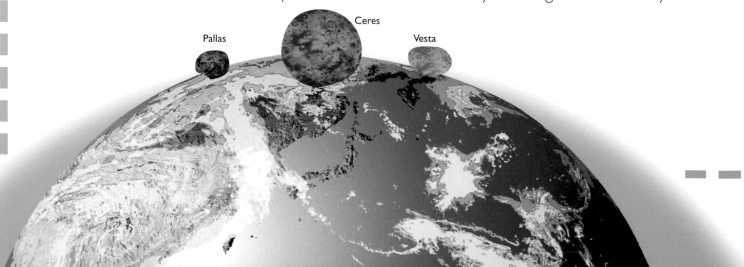

between the orbits of Mars and Jupiter. He named it Ceres. Then, in 1802, German astronomer Heinrich W. M. Olbers noticed another bright object also between the orbits of Mars and Jupiter. Like Piazzi, he first believed the object was a comet, but it turned out to be another kind of small body. Olbers named the orbiting object Pallas. He used the word *asteroid,* which means "starlike," to describe his discovery. Over the next four years, two more asteroids, Juno and Vesta, were discovered orbiting between Jupiter and Mars. Since that time, thousands of asteroids have been found.

THE BIG THREE

Scientists have discovered at least 26 asteroids that are larger than 125 miles (200 km) in diameter. The rest range in size from less than 125 miles to just over ½ mile (1 km) across. The largest three asteroids are Ceres, Pallas, and Vesta. Ceres, by far the largest, has a diameter of about 600 miles (950 km). Pallas and Vesta are both about 340 miles (550 km) in diameter. Scientists believe that these three large asteroids are nearly ball shaped. But most asteroids have an irregular shape, sort of like a potato. Others appear to be longer and narrower.

ASTEROIDS DARK AND BRIGHT

Asteroids are divided into three main groups according to their makeup. Most asteroids are dark, stony rocks called C-types. These asteroids are difficult to see because they are dull in color and do not reflect much of the Sun's light. A second group, the S-type asteroids, are brighter than the C-types and tend to be easier to see. They are gray in color and contain some metal. The rarest asteroids—the M-types—are also the brightest. They are made of pure nickel-iron metal.

Left: Giuseppe Piazzi (1746-1826) discoved the first asteroid, Ceres, in 1801.

Below: This color map of the asteroid Vesta shows the elevation, or height, of different parts of its surface. The bright colors represent the highest regions.

Elevation

-7 Miles
(-11½ Km)

7 Miles
(11½ Km)

Amor

main asteroid belt

Eros

Ceres

Sun

Apollo

Hidalgo

Adonis

Above: Most asteroids, including the largest, Ceres, orbit within the asteroid belt. But some asteroids have orbits that take them inside or outside the belt.

ASTEROID ORBITS

Most asteroids orbit the Sun between Mars and Jupiter in an area called the asteroid belt. The middle of the belt lies about 250,000,000 miles (400,000,000 km) from the Sun. Occasionally, asteroids are knocked out of the asteroid belt when they collide with one another. Others may be tugged off course by the gravity, or pull, of a larger body, such as Jupiter. These asteroids are tugged away from the asteroid belt into the outer solar system or pulled closer to Earth into the inner solar system.

A number of the asteroids orbiting in the inner solar system cross paths with Earth's orbit. Astronomers call these Near-Earth asteroids. A few have come closer than 100,000 miles (160,000 km) to Earth. This puts them dangerously near our planet. Although the chances are quite small, astronomers believe that an asteroid could collide with Earth in the future and cause enormous damage.

THE TROJANS

In addition to the asteroids in the asteroid belt and those that cross Earth's orbit, there are two small groups of asteroids that travel in the same orbit as Jupiter. They are known as the Trojans. One group travels in front of Jupiter; the other travels behind the planet. Astronomers know of several hundred Trojan asteroids, but they believe that 1,000 or more may exist.

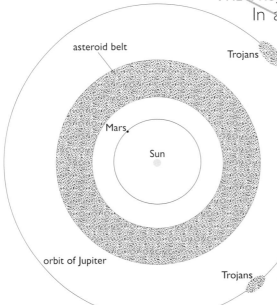

asteroid belt

Trojans

Mars

Sun

Jupiter

orbit of Jupiter

Trojans

Left: The two groups of Trojan asteroids have been captured by Jupiter's powerful gravity and circle the Sun in Jupiter's orbit.

MOONS OR ASTEROIDS?

Most asteroids orbit the Sun, but some asteroids may orbit certain planets as moons. Some astronomers think that Mars's two small moons, Phobos and Deimos, were once asteroids from the nearby asteroid belt. They might have been captured by the pull of Mars's gravity when they came close to the planet billions of years ago. Some of the small outer moons of Jupiter, Saturn, Uranus, and Neptune may once have been asteroids, too.

Ceres weighs about one-third as much as all the asteroids put together.

The 10 Biggest Asteroids			
Name	Year discovered	Diameter (miles)	(km)
Ceres	1801	580	940
Vesta	1807	340	550
Pallas	1802	325	525
Hygiea	1849	280	450
Euphrosyne	1854	230	370
Interamnia	1910	220	350
Davida	1903	200	350
Europa	1858	195	315
Cybele	1861	190	310
Juno	1804	155	250

Phobos, the largest of Mars's two moons, is probably a captured asteroid. It measures less than 20 miles (30 km) across.

Our Moon

Some scientists believe that an asteroid may have created our Moon. Billions of years ago, a very large asteroid may have crashed into Earth and knocked off large rocky pieces of the planet. These rocky chunks were flung into space. In time, they may have come together to form a separate body, the Moon.

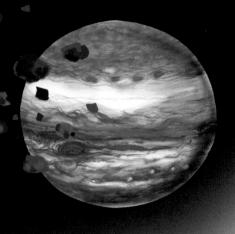

Left: The leftover lumps of rock and metal between Mars and Jupiter never formed into another planet. Instead, they continued to orbit the Sun as asteroids.

WHERE ASTEROIDS CAME FROM

For many years, astronomers thought that the asteroids were the remains of another planet. This planet began forming between Mars and Jupiter, they said, at the same time as the other planets were forming. Then it gradually moved closer and closer to Jupiter. Jupiter's very strong gravity put a great deal of pressure on this planet. The gravity created powerful tides inside the planet, which eventually pulled it apart into pieces. Scientists believed that these pieces became the asteroids.

Most astronomers no longer believe that the asteroids are pieces from an exploded planet. Instead, they think that the asteroids are leftover lumps of matter in the solar system that never formed into one larger body.

Life of an asteroid:
1. Small pieces of rock and metal fuse together to form an asteroid.
2. The asteroid melts, and any metal it contains sinks to the center.
3. A cross-section of the newly formed asteroid shows the solid iron core and the rocky outer layer.

1.

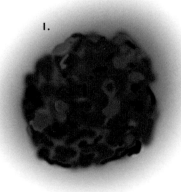

2.

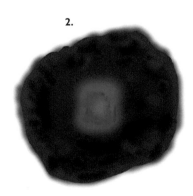

3.

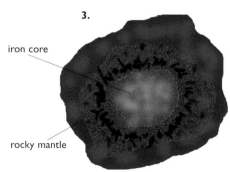

iron core

rocky mantle

In the early days of the solar system, a disk of matter containing gas, dust, and small lumps of rock and metal circled the newborn Sun. Within this disk, the lumps kept colliding and sticking together, gradually forming larger and larger bodies. In time, these bodies formed into the planets. But the lumps of matter between Mars and Jupiter never came together as a planet. When these lumps of rock and metal collided together, Jupiter's powerful gravity pulled them apart.

LIFE STORY

These newly formed asteroids were made up of rock, dust, and metal bound together. Over time, some of the bigger asteroids probably melted. At this stage, the heavy metal in the large asteroids sank to the center to form a metal core. The lighter rock formed layers around the core, and the whole asteroid slowly cooled and became solid.

But few asteroids have remained as they were during their early formation. Collisions among asteroids broke them up into smaller pieces and sometimes shattered them completely. In some cases, when a big asteroid shattered from a powerful collision, its metal core was exposed, creating a rare M-type asteroid. In other cases, the rocky crust was not entirely broken away after an asteroid collision, resulting in the more common rocky types of asteroids.

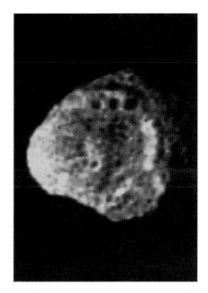

Above: This distant moon of Saturn is probably a captured asteroid. Its craters show that it has been battered repeatedly in collisions with other asteroids.

The most violent collisions between asteroids can smash them to pieces and produce hundreds of smaller fragments.

Most comets are named after the person or persons who discover them.

Inside a comet's bright coma, or head, is a tiny nucleus. Well-developed comets have two tails. The bright one is made up of dust, while the faint one consists of glowing gas.

Comets

Comets are a mixture of rock and dust bound together with ice and frozen gases. They visit our skies regularly, sometimes shining more brightly than the stars.

Comets can be a spectacular sight. At their brightest, they are easily seen with the naked eye for weeks or months at a time. In ancient times, the sudden appearance of a comet terrified many people. They believed a comet was a bad sign and that wars, disease, and other disasters would follow.

It is easy to see why comets impressed ancient peoples. Comets can look extremely large—their tails can appear to stretch halfway across the sky. A comet's brightest parts blot out the light of hundreds of stars. However, comets appear to be much bigger than they are. A comet's solid core is often only several tens of miles in diameter, smaller than most bodies in the solar system.

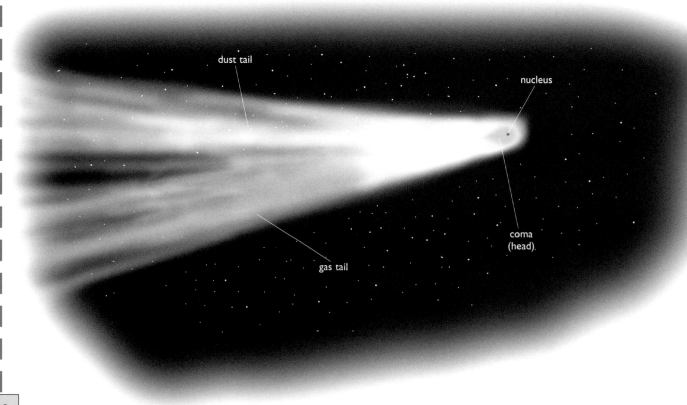

dust tail

nucleus

gas tail

coma
(head)

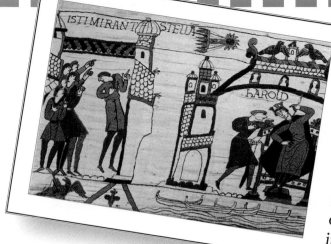

Harold's Comet

Ancient Britons believed they had good reason to fear the comet that appeared in 1066. In that same year, Britain was invaded by the French. The French killed the British ruler, King Harold, and conquered the Britons. The comet appears in the Bayeux Tapestry, a large embroidered cloth that illustrates the invasion of ancient Britain.

LOOKING AT COMETS

A comet has three main parts—the nucleus, the head, and the tail. The two visible parts are the head and the tail. The head, which astronomers call the coma, is the brightest part. It can measure tens of thousands of miles across. Deep in the core of the coma is the only solid part of the comet, the nucleus. The nucleus is small, sometimes measuring less than 10 miles (16 km) across.

Streaming away from the head is the comet's tail, which can grow to enormous lengths of millions of miles. Many comets grow two tails, one straight and the other curved.

Bennett's Comet was the brightest comet of 1970. At the time this picture was taken, only the dust tail was clearly visible.

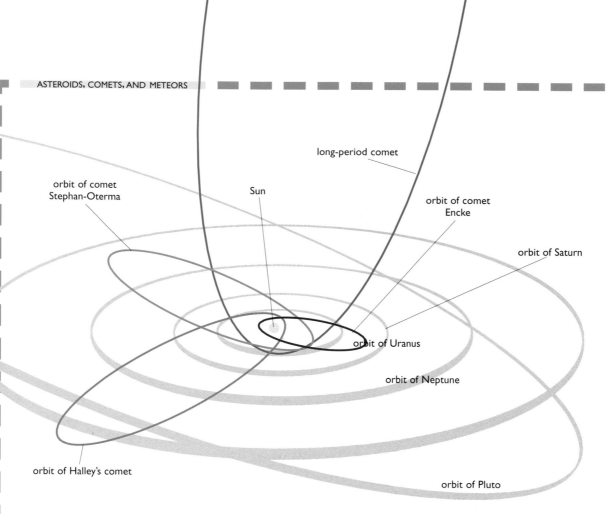

long-period comet

orbit of comet
Stephan-Oterma

Sun

orbit of comet
Encke

orbit of Saturn

orbit of Uranus

orbit of Neptune

orbit of Halley's comet

orbit of Pluto

STAR POINT

Encke's Comet has
the shortest known
orbit of any comet. It
orbits the Sun in just
3.3 years.

ORBITING THE SUN

Like the planets, comets travel through space in an orbit that takes them around the Sun. Comets that take less than 200 years to complete their orbit are called short-period comets. They are regular visitors to our skies. The best known short-period comet is Halley's comet, which returns to our skies about every 76 years.

Other comets follow extremely long orbits. Their journeys around the Sun can take thousands or even millions of years. A comet that takes more than 200 years to orbit the Sun is called a long-period comet. Long-period comets travel to distant reaches of the solar system before heading back in toward the Sun.

STARTING TO SHINE

During most of its orbit, a comet is invisible to us. It has no coma or tail, and its small nucleus cannot be seen from Earth, even with powerful telescopes.

But as an icy comet gets within a few hundred million

Comet Kohoutek

In 1973, a long-period comet named Kohoutek was first spotted in telescopes when it was more than 430 million miles (700 million km) from the Sun. Comets cannot usually be detected at such a distance, so astronomers predicted that it was going to be the brightest comet of the century. But when the comet did approach the Sun, scientists were surprised to find that it was barely visible to the naked eye. Astronauts in the U.S. space station *Skylab* studied this comet, making it the first comet to be studied closely from space. Kohoutek will not return to Earth's skies for at least another 75,000 years.

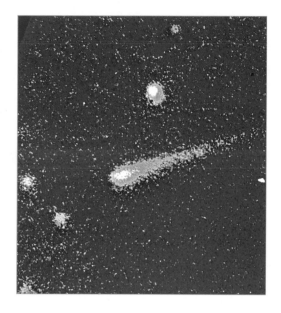

comet nucleus

beginning of cloud

cloud expands

A cloud of gas and dust gradually grows around the nucleus of a comet as it approaches the Sun. As the cloud reflects sunlight, it becomes visible in our skies.

comet travels toward the Sun

miles of the Sun, things start to change. The Sun's heat warms the outer surface of the comet and melts some of its ice. The ice changes into gas and forms a cloud around the nucleus. At the same time, the dust that was frozen in the ice is released in the gas. This cloud of gas and dust begins to reflect sunlight. As the comet comes closer to the Sun, the cloud around the nucleus grows larger and shines more brightly in the sunlight.

cloud starts to be pushed away from nucleus

sunlight

The tail of a comet is longest when it is closest to the Sun, and it always points away from the Sun.

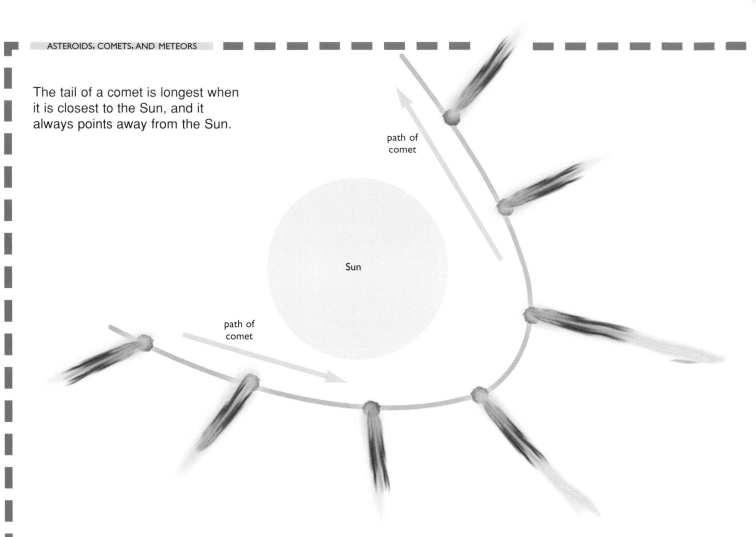

path of
comet

Sun

path of
comet

SUN PRESSURE

The Sun also has another effect on the cloud of gas and dust around the comet. Pressure from the Sun's radiation and the solar wind (a fast-moving stream of particles that move outward from the Sun) gently press against the cloud.

The Sun's radiation and the solar wind push against the particles of gas and dust around a comet's nucleus, forming the comet's tail.

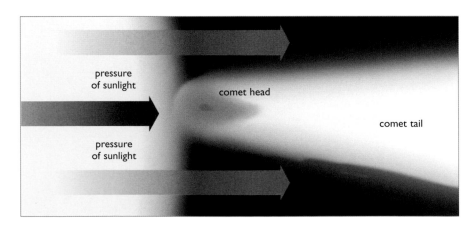

pressure
of sunlight

comet head

comet tail

pressure
of sunlight

These forces push some of the cloud's dust and gas away from the main body of the comet. Over time, a stream of dust and gas forms the comet's tail.

The tail always points away from the Sun, so when the comet orbits toward the Sun, its tail follows behind. But when the comet moves away from the Sun in its orbit, its tail streams out in front of it.

INTO THE DEEP FREEZE

As the comet gets farther away from the Sun, it grows colder. Not as much ice melts, and not as much gas and dust are released. The head and tail start to shrink and gradually fade. After a while, the comet freezes completely. There is no cloud to reflect sunlight. Once again, the comet becomes invisible to us. It will not be seen again until it next closes in on the Sun—in some cases, not for thousands of years.

WHERE COMETS CAME FROM

Comets are ancient remains of ice and dust left over after the planets formed. Astronomers think that most comets collected in a "cloud" at the outer reaches of the solar system, where billions of them remain. This comet storehouse is called the Oort Cloud, and it surrounds the solar system. Many of the comets that appear in our skies come from the Oort Cloud, but others come from an area closer to the Sun, called the Kuiper Belt. This belt is just beyond Neptune's orbit, near Pluto.

Comet West, the brightest comet of 1976, was easily visible to the naked eye. Astronomer Richard West had discovered the comet the previous year.

STAR POINT

Every time a comet travels around the Sun, it can lose as much as 500 million tons of material.

FAMOUS COMET

Halley's comet, also called Comet Halley, is the most famous comet. People have spotted it every time it has returned to Earth's skies since 240 B.C. It was the comet that appeared at the time of the Battle of Hastings in A.D.1066. It last appeared in 1986, but it was not easy to see because it was so faint.

The comet is named after the English astronomer Edmond Halley. He saw it in 1682 and, after checking past records, realized it was similar to comets seen in 1531 and 1607. He guessed that it was the same comet, and he predicted that the comet would return again 76 years later, in 1758. It did, 16 years after Halley had died.

At its last appearance, in 1986, Halley's comet was a faint object in the night sky.

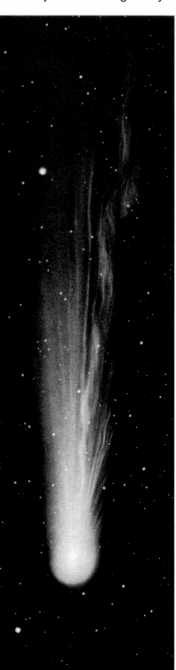

This ancient woodcut shows the bright comet of 1456, which we now know was an appearance of Halley's comet.

Meteoroids, Meteors, and Meteorites

Meteors appear as streaks of light in the night sky. Occasionally, they fall through our skies and land on Earth's surface as meteorites.

Outer space is full of specks of rock and metal. These small pieces of material are called meteoroids. Like asteroids and comets, meteoroids orbit the Sun.

On its own journey through space, Earth comes across these orbiting pieces all the time. It attracts nearby meteoroids with the pull of its gravity. These tiny specks of matter leave bright trails behind them as they travel through Earth's atmosphere, or the layer of gases around the planet. From Earth, these bright trails look like stars falling from the sky. On a clear night, you might be able to see several "falling stars" every hour. Scientists call these trails of light meteors.

This woodcut from 1557 illustrates a meteor shower as a battle in the heavens.

A meteor burns a trail through Earth's atmosphere. The flash near the end of the trail shows where the meteor has exploded.

BURNING UP

As a meteoroid falls toward Earth, it enters Earth's atmosphere. The friction, or rubbing, of Earth's atmosphere against the surface of the falling particle causes it to heat up. As it gets hotter and hotter, it starts to burn. When a meteoroid leaves outer space and begins to burn through Earth's atmosphere, it is called a meteor. As it burns, the meteor leaves a flaming trail behind it, until it burns away to nothing. A meteor's bright trail is often easy to see in our night sky.

The burning trails of most meteors last only a few seconds. Many meteors are no bigger than a grain of sand, and their small size causes them to burn up quickly. Occasionally, a larger meteor enters Earth's atmosphere. It

Some meteors burn up completely as they plunge through the atmosphere. A few are big enough to survive and reach the ground.

meteor burns off in upper atmosphere

meteor leaves bright trail

meteor breaks up

huge meteorite falls to Earth, causing a crater

burns brighter and for a longer period of time. An especially bright meteor is called a fireball.

SHOWERS OF METEORS

At certain times of the year, many meteors appear to rain down at once in the night sky. Scientists call this increased meteor activity a meteor shower. Meteor showers occur when Earth passes through tiny pieces of rock and metal left behind by an orbiting comet. These clusters of leftover pieces are drawn into Earth's atmosphere, where they burn up and streak through our skies.

Regular showers take place every year in a certain part of the sky. They are named after the constellation, or pattern of stars, that the meteors appear to come from. The Perseids, which occur in August, appear to come from the constellation Perseus. The Orionids, in October, seem to come from the constellation Orion. Earth passes through these clusters of comet particles around the same time each year, so astronomers can predict when the meteor showers will occur.

STAR POINT

Meteor particles hit Earth's atmosphere at speeds of up to 150,000 miles (250,000 km) an hour.

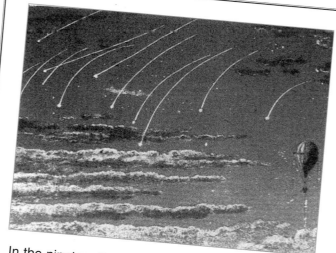

In the nineteenth century, scientists sometimes used hot air balloons to get a better view of events in the night sky. This Leonid shower was observed in 1870.

Meteor Storm

The Leonids, a meteor shower that appears to come from the constellation Leo, occurs each year in November. Chinese astronomers saw the first Leonid meteor shower in A.D. 902. They thought the stars were falling like rain. The 1966 Leonid shower was one of the greatest meteor showers on record. So many meteors fell, it was called a meteor storm. At one point during the storm, some observers saw 40 meteors each second!

Above: This is the world's biggest meteorite, found in southern Africa. It fell to Earth in prehistoric times.

METEORITES

Sometimes Earth crosses paths with a much larger meteoroid than usual. The meteoroid drops through the atmosphere and starts burning as a meteor. But if it is big enough, only its outer layer will burn away. The rest of the meteor will fall to the ground. When a meteor has fallen through Earth's atmosphere and lands on Earth, it is called a meteorite. Meteorites weighing thousands of pounds have been found all over the world. The biggest, the Hoba meteorite, was found in Namibia in southern Africa. It weighs about 120,000 pounds (54,000 kg).

There are three main kinds of meteorites. The largest group, stony meteorites, are made up mainly of rocky materials. They account for over 90 percent of all meteorites that land on Earth. A second kind are made of iron and are called iron meteorites. About 1 percent of meteorites contain nearly equal amounts of stone and metal. They are called stony-iron meteorites.

These two meteorites were found in Antarctica. Astronomers think that one **(above)** was probably broken off from Mars and the other **(right)** was broken off from the Moon.

WHERE METEORS CAME FROM

Where did these specks and lumps of rock and metal come from? Some meteors and meteorites are probably pieces of the Moon and other planets, such as Mars. Smaller pieces from these larger bodies were broken off in collisions with asteroids. Other meteors and meteorites probably came from the asteroid belt. They formed when asteroids slammed together and broke apart into smaller pieces. Still others are probably dust particles left behind by comets.

DIGGING CRATERS

Most meteorites are slowed down as they travel through Earth's atmosphere. By the time they reach Earth's surface, their speed and size are not great enough to hit Earth with much force. Only the largest meteorites dig out craters, or large pits, when they hit Earth's surface.

Scientists believe there are over 100 meteor craters on Earth. In the past there were many more. But over millions of years, ancient craters have been worn away by the process of erosion. Erosion is the gradual wearing away of Earth's surface by natural elements such as wind or flowing rivers.

STAR POINT

The largest meteorite in a museum is the Ahnighito, located in the American Museum of Natural History in New York. Explorer Robert Peary found it in Greenland in 1897.

This map shows the locations of known craters made by meteorites.

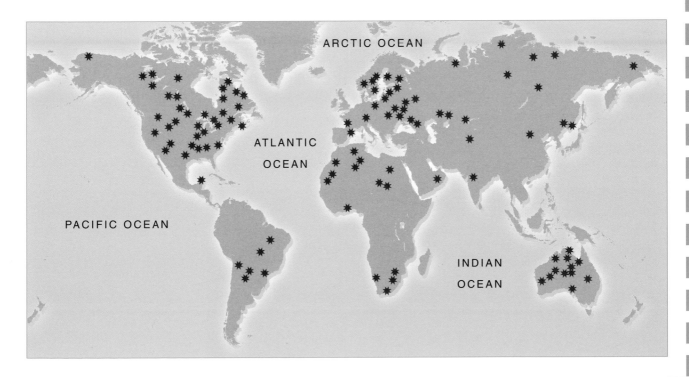

Barringer Crater in the Arizona desert is the most famous and best-preserved meteorite crater on Earth.

Mercury's surface is covered with craters. It was battered by meteorites billions of years ago.

THE WORLD'S BIGGEST CRATER

The best-preserved and largest meteorite crater on Earth is the Barringer Crater, also called Meteor Crater, in Arizona. It is a great pit in the Canyon Diablo region of the Arizona desert, located about 20 miles (32 km) west of the town of Winslow. It measures more than 4,000 feet (1,200 m) across and about 600 feet (180 m) deep.

The crater is named after Daniel Barringer, a silver miner in the early 1900s. He was one of the first people to be

convinced that a meteorite had formed the crater. He drilled into the crater floor and found a layer containing pieces of nickel. The nickel pieces were remains of the meteorite that had blasted out the crater.

Scientists believe that this meteorite fell to Earth about 50,000 years ago. It would have been the size of a railroad car and weighed as much as 300,000 tons. No large pieces of the meteorite have ever been found. But this is not surprising. The meteorite would have hit the ground so hard that it would have smashed to pieces. All that remains of the monster meteorite are many pieces of iron found in the surrounding region.

SHAPING PLANETS AND MOONS

Many other planets and moons in the solar system have also been shaped by collisions with meteorites. Most collisions took place billions of years ago, not long after the planets and their moons formed.

You can see the result of these ancient collisions if you look at the Moon. Much of its surface is covered with craters large and small. The craters have barely changed since they were formed. Because the Moon does not have any atmosphere to create weather, there has been no erosion to wear away the craters. Among the planets, Mercury is the most heavily cratered body. Like the Moon, it has never had an atmosphere, so there has been no erosion to change its surface. Venus and Mars have fewer craters because, like Earth, both planets have an atmosphere.

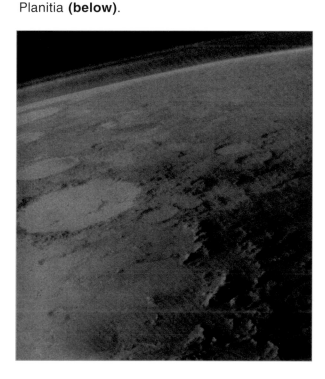

Meteorites created most of the craters on the Moon **(above)**. A meteorite impact made the large crater on Mars named Argyre Planitia **(below)**.

Visiting Comets and Asteroids

For years, astronomers had only been able to study comets and asteroids from far away. Space probes have given us our first glimpses of these small worlds from close up.

In 1986, Halley's comet returned to Earth's skies for the first time since 1910. As the comet passed closest to Earth in March 1986, five space probes were sent to study it. They were launched by Russia, Japan, and Europe. The probes were part of a worldwide investigation known as International Halley Watch, in which over 800 scientists in 40 countries took part.

Two Russian probes, *Vega 1* and *Vega 2*, were launched in December 1984. In January 1985, the first of two Japanese probes was launched. It was named *Sakigake*, meaning pioneer. This was a good name because *Sakigake* was the first probe Japanese space scientists had ever launched. In July, the European Space Agency (ESA) launched its probe, *Giotto*. Japan's second probe, *Suisei*, meaning comet, followed in August.

Above: *Vega 1* and *Vega 2*, similar to the spacecraft above, were the first probes to study Halley's comet.

Right: The *Suisei* spacecraft was small compared to the *Vega*. It weighed only about 300 pounds (140 kg), compared to *Vega*'s 5 tons.

THE ENCOUNTERS

The two Russian probes, *Vega 1* and *Vega 2*, were the first to encounter, or meet, Halley's comet. They took pictures, measured the amount of dust around the comet, and investigated the contents of the comet's nucleus.

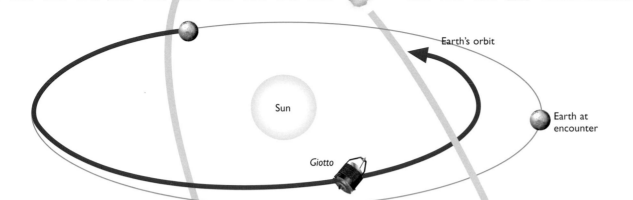

Halley's comet

Earth's orbit

Sun

Earth at encounter

Giotto

path of Halley's comet

Giotto took eight months to travel to Halley's comet. It suffered some damage when it was bombarded by comet particles but went on to encounter another comet, Grigg-Skjellerup, in 1992.

Giotto

Scientists learned that Halley's comet is made up mainly of frozen water and frozen carbon dioxide, or dry ice.

The two Japanese probes, *Suisei* and *Sakigake*, made useful observations on the effect of the comet on the solar wind, the stream of particles coming from the Sun. And *Suisei* took pictures of bright gas jets coming from the comet.

Europe's *Giotto* was the last probe to encounter Halley. It flew past the comet at a distance of only about 370 miles (600 km), the closest of all the probes. It took pictures that showed the bright gas jets and also the comet's nucleus. The nucleus was dark in color and only about 10 miles (16 km) across.

Giotto snapped this picture of gas jets shooting out from the nucleus of Halley's comet.

Above: *Galileo*'s on-board rocket engine fires after the probe has been placed in orbit by space shuttle *Atlantis* on October 18, 1989.

VISITING ASTEROIDS

In October 1989, *Galileo* was successfully launched by the National Aeronautics and Space Administration (NASA). *Galileo* left Earth on a roundabout route that would eventually take it to Jupiter. NASA scientists planned for it to pass twice through the asteroid belt and take close-up pictures of asteroids for the very first time.

Galileo followed a long looping path through the solar system. It first flew near Venus, and it was speeded up by the pull of Venus's gravity. Then it flew back to Earth, where it was again speeded up by Earth's gravity. It then sped out toward the asteroid belt. In October 1991, *Galileo* turned its instruments on an asteroid named Gaspra, which proved to be irregularly shaped and heavily cratered.

Galileo next looped back toward Earth to gather even more speed before returning to the asteroid belt. In August 1993, its target was a larger asteroid called Ida. Scientists were amazed to discover that Ida had a tiny moon circling around it. They named the moon Dactyl.

Next *Galileo* continued through the asteroid belt to its main target, Jupiter, where it would explore the giant planet's atmosphere and moons.

Left: This lump of rock is the asteroid Ida. *Galileo* photographed the asteroid in 1993 on its way to Jupiter. Ida measures about 35 miles (55 km) long.

Bombarding Earth

Huge chunks of rock from outer space have bombarded Earth in the past and may do so again in the future.

If an object the size of the meteorite that created Barringer Crater hit New York City, the city would be flattened. Much larger meteorites and asteroids have hit Earth in the more distant past. They have not only torn out great craters, they have also affected weather conditions and wiped out many plant and animal species.

A collision with a huge asteroid may have been the reason dinosaurs died out about 65 million years ago. Some scientists suggest that when this large asteroid hit Earth, it blasted millions of tons of rock and dust into the air. These particles quickly spread through the atmosphere, covering Earth with a dust cloud. This layer of dust was so thick and black that light from the Sun could not get through. Without sunlight, plants would have stopped growing, because plants need sunlight to make their food. Since most animals rely on plants for food, many of them, including the dinosaurs, would have died out.

Comet Collisions

Not all collisions with objects from outer space have caused destruction on Earth. Some might have helped bring about life. Billions of years ago, Earth did not have all of the elements needed to produce and maintain life. It is possible that icy comets crashing into Earth brought water and other raw materials to our planet. Over time, these materials may have helped make it possible for life to exist on Earth.

New York City would be completely destroyed if a large asteroid smashed into it.

THE GOOD NEWS

Some Near-Earth asteroids regularly pass a few hundred thousand miles away from Earth. This may seem far away, but it's a short distance in outer space. Just a slight change in an asteroid's orbit could direct it toward Earth. Because of this danger, several groups of astronomers around the world closely watch Near-Earth asteroids.

Fortunately, astronomers think that the chances of large meteorites or asteroids hitting Earth are extremely small.

CHANGING COURSE

If one day we find that there is a large asteroid heading for Earth, what could we do? Scientists have suggested several ideas. One is to send spacecraft with powerful rocket engines to land on the asteroid and use the force of the engines to cause the asteroid to change course. Another idea is to blast an approaching asteroid apart with powerful bombs. But since these ideas have never been tested, no one knows if they would work.

KEEPING WATCH

Astronomers will continue to explore and investigate the small worlds that orbit the Sun. The more we learn about these tiny, ancient worlds, the more we can understand how the solar system began and developed nearly 5 billion years ago.

Some scientists believe that rockets carrying powerful nuclear bombs might protect Earth from an approaching asteroid. The hopes are that if the bombs exploded close to the asteroid in outer space, the blast would break the asteroid apart.

Some astronomers use observatories like this one, located in Hawaii, to watch out for asteroids that may be heading dangerously close to Earth.

Glossary

asteroid: a small rocky body that circles the Sun or another heavenly body

asteroid belt: the region in the solar system between the orbits of Mars and Jupiter, where most asteroids are found

atmosphere: the layer of gases around Earth or another heavenly body

coma: the shining head of a comet, which surrounds the nucleus

comet: a small body in the solar system, made up of dust and ice, which shines when it approaches the Sun

constellation: a pattern of stars in the sky

core: the center part of a planet, moon, or other body

crater: a pit on the surface of a planet, moon, or other body

erosion: the gradual wearing away of the landscape by the weather, moving water, and other forces in nature

gravity: the attraction, or pull, that every heavenly body has on things on or near it

meteor: a streak of light produced when a meteoroid burns up in Earth's atmosphere

meteorite: a lump of rock or metal from outer space that falls to the surface of a planet, moon, or other body

meteor shower: a group of meteors that streak through a part of the sky during a certain time of the year

meteoroid: a speck or small lump of rock or metal orbiting the Sun

Near-Earth asteroid: an asteroid with an orbit that may take it dangerously close to Earth

nucleus: the solid part of a comet

orbit: the path in space of one heavenly body around another, such as an asteroid around the Sun

period: the length of time a heavenly body takes to complete its orbit around another body

planet: a large body that orbits the Sun

probe: a spacecraft that travels from Earth to explore bodies in the solar system

solar system: the Sun and all the bodies that circle around it, including the planets, asteroids, comets, and meteorites

solar wind: a fast-moving stream of particles that travels outward from the Sun

Trojans: two groups of asteroids that travel in the same orbit as Jupiter

Index